PENGUINS

PENGUINS

TOM JACKSON

This pocket edition first published in 2024

First published in a hardback edition in 2020

Copyright © 2024 Amber Books Ltd

All rights reserved. No part of this publication may be reproduced, stored in a retrieval system, or transmitted in any form or by any means, electronic, mechanical, photocopying, recording, or otherwise, without prior written permission of the copyright holder.

Published by
Amber Books Ltd
United House
London N7 9DP
United Kingdom

www.amberbooks.co.uk
Instagram: amberbooksltd
Pinterest: amberbooksltd
Twitter: @amberbooks

ISBN: 978-1-83886-360-9

Project Editor: Anna Brownbridge
Designer: Keren Harragan and Rick Fawcett
Picture Research: Terry Forshaw

Printed in China

Contents

Introduction   6

Types of Penguin   8

Penguins in Water   98

Penguins on Land   134

Penguin Family   168

Penguin Chicks   194

Picture Credits   224

Introduction

There is something endearing, certainly, and perhaps comical, about penguins, but also plenty to admire. No other kind of bird is anything like the penguins; none come close. Dressed in a smart suit of black-and-white feathers, these birds waddle and wag in a somewhat inelegant fashion on land, but glide with ease through water, leaving a glittering trail of bubbles.

The story of penguins is the story of survival in some of the  toughest habitats on Earth. Frequently associated with the frozen world of the Antarctic, penguins occupy a wide range of ocean habitats, from the icy ocean to the rocky coasts of desert landscapes.

All the while, penguins manage to reflect something that we all recognise: a drive above all else to care for their young, to cooperate within a wider society – and enjoy a good meal when time allows.

ABOVE:
It is the male emperor penguin that cares for a chick as it emerges from the shell, feeding it its first meal and protecting it from the cold.

OPPOSITE:
A pair of magellanic penguins show off the features that allow these plucky seabirds to survive on the ice and in the ocean.

Types of Penguin

The number of penguin species is in flux, as genetic analysis of remote communities suggests to some authorities – although not others – that these groups constitute new species in their own right. The species number currently stands at between 16 and 21, and all but one is found in the Southern Hemisphere. Only the Galápagos penguin, which occupies the Pacific islands of that name that span the equator off the coast of Ecuador, sneak into the Northern Hemisphere. And while the majority of the species are polar birds, ones that are at home on the sea ice and frozen seas of the Southern Ocean, the first to be inspected and recorded by scientists were from more temperate locations. The first record of a penguin comes from the Portuguese explorer Vasco da Gama, who came across the little birds around the Cape of Good Hope in the 1490s.

Quite how the birds took the name penguin is open to debate. It might be that the name comes from the Latin term pinguis, meaning "fat" – archaic names in other languages are based on a similar reasoning. An alternative is that the word has a Welsh root, and means "white head". The sharp-eyed reader will soon see that few penguins have white heads (quite the opposite), so it is likely that "penguin" originally referred to great auks. These flightless northern seabirds, related to puffins but now extinct, had more than a passing resemblance to penguins. However, despite appearances, penguins are actually more closely related to albatrosses and petrels.

OPPOSITE:
Gentoo Penguin, Antarctica
Built for cold air and cold water, this robust bird has 15 feathers on every square centimetre (70 on every square inch) of its body.

King Penguins, Falkland Islands
The second largest penguin species, the king penguin lives mostly on the sub-Antarctic islands, although a few colonies form on the extreme south of the South American mainland.

OPPOSITE:
King Penguin, South Georgia
The stark black-and-white colouring helps the bird stay hidden in water, blending with the dark water when viewed from above, and the pale sky when seen from underneath. Any variation makes them vulnerable to attack.

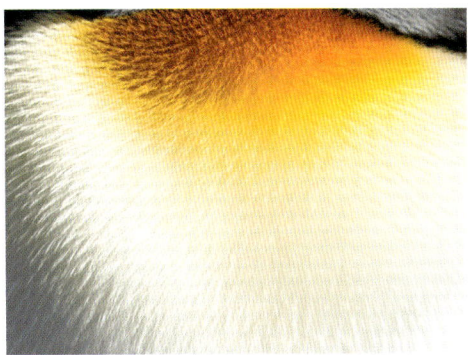

TOP LEFT:
King Penguins, South Georgia
Unlike most species, king penguins do not mate for life, but find a new partner each spring with elaborate courtship rituals. The annual reproductive cycle begins at the beginning of the Antarctic winter, in March or April.

MIDDLE AND BOTTOM LEFT:
King Penguin, Falkland Islands
The penguins select mates based on the vibrancy of their plumage. The coloured feathers on the chest and bill reflect ultraviolet light, which is highly visible and serves as a signal of a healthy immune system.

King Penguins, East Falkland Island
Males of this species are slightly larger than the females, withboth sexes ranging between 85 and 95cm (33–37in) tall.

LEFT:
African Penguin, South Africa
The only African species is also called the jackass penguin because its braying call sounds similar to that of a donkey. Also sometimes referred to as the black-footed penguin, this species grows to about 60cm (24in) tall.

ABOVE:
African Penguin
The home territory of this species seems at odds with what is familiar about penguins. On land their rookeries and breeding grounds are close to dry meadows and semideserts, and even the Namib, one of the driest desert habitats on Earth. However, the water off this coast comes chilled straight from Antarctica, creating an ocean ecosystem teeming with fish.

RIGHT:
African Penguin
The African species was under threat of extinction in the 1970s and 1980s due to egg collection and pollution from the South African Cape's shipping lanes. Today its island colonies are better protected.

OVERLEAF:
Erect-crested Penguins
Erect-crested penguins use the Antipodes and Bounty Islands south of New Zealand to breed, and to a lesser extent other sub-Antarctic islands in the region.

LEFT:
Erect-crested Penguins, Antipodes Islands
These 65cm (26in) tall penguins spend the whole winter out at sea. The summer breeding ground is barren and treeless. The birds weigh between 2.5kg and 6kg (5.5–13.2lbs). As with other penguin species, the male is usually slightly larger than the female.

ABOVE:
Erect-crested Penguin
Like its relatives, the rockhopper and macaroni penguins, erect-crested penguins have a yellow stripe running from the eye to a feathery tuft. This species is distinctive in that the tuft sticks straight up.

LEFT:

Gentoo Penguins
A small number of penguins, about one in 50,000, are isabelline, they have a pale grey colour, not the usual blue-black body parts.

BELOW:

Gentoo Penguins, Falkland Islands
A group – known as a "waddle" – of penguins returns to land after foraging. They live on islands in the Atlantic and Pacific zones of the Southern Ocean.

LEFT:
Gentoo Penguins, Antarctic Peninsula
Gentoo penguins approach the water. As well as being based on sub-Antarctic islands, about ten percent of live on the Antarctic mainland and its sea ice.

ABOVE TOP:
Gentoo Penguins, Falkland Islands
The Falklands has the largest population of gentoo penguins. They are attracted to the beaches as they are mostly ice free, and surrounded by the shallow seas of the continental shelf, with more food than deeper waters.

ABOVE BOTTOM:
Gentoo Penguin, Falkland Islands
A male attracts females by making a bellowing call. Once they bond, they will search for each other for many years. They will collect stones from the beach one at a time with their beaks to construct a round nest.

PREVIOUS PAGES:
Little Penguins, Australia
This species digs its own burrows into sandbanks near the shore, or takes refuge in crevices in rocks, as shown here.

LEFT:
Little Penguin
This species lives on the southern coast of Australia and parts of New Zealand. Standing less than 50cm (20in) tall, the little penguin is so named because it is the smallest species of penguin.

RIGHT:
Yellow-eyed Penguin
This unusual species lives on the southern coast of New Zealand's South Island and several islands further out in the Southern Ocean. It is the largest penguin species in and around New Zealand. It is thought this species spread to the mainland around 1500, after other species were driven to extinction through hunting by the first people to settle in New Zealand.

BELOW TOP:
Yellow-eyed Penguin, South Island, New Zealand
Biologists think that this species is something close to the first penguins around 25 million years ago.

BELOW BOTTOM:
Yellow-eyed Penguin, Dunedin, New Zealand
This is one of the most endangered species, with an estimated 4000 individuals in the wild. It is threatened by habitat loss and mammal predators.

Emperor Penguins, Antarctica
This is the only penguin species – in fact the only large animal of any kind – that stays on the Antarctic continent during the winter. Other penguins move out to sea to avoid the harshest weather.

OPPOSITE:
Emperor Penguins, Antarctica
The penguins sit out the winter to incubate their eggs, which can only develop on land. The chicks inside would freeze without their parents' help to keep them warm.

TOP LEFT:
Emperor Penguins, Antarctica
The penguins help each other to stay warm by huddling together in temperatures that go as low as −70°C (−95°F).

BOTTOM LEFT:
Emperor Penguins, Antarctica
As the tallest and heaviest penguin species, the emperor penguin reaches 115cm (45in) tall. Emperor penguins typically weigh from 22–45kg (49–99lbs).

Emperor Penguin, Antarctica
In common with its close relative, the king penguin, the emperor penguin is easily identified by the bright colouring around the ears, although the chest plumage is more muted than its relative.

PREVIOUS PAGES:
Humboldt Penguins
These banded penguins are distinguished from their closest cousins, the Magellanic penguins, by the single dark band on the upper chest.

LEFT:
Humboldt Penguin
This medium-sized species lives along the coasts of Chile and Peru, where the cold-water Humboldt Current glides northward. Its rutted beak channels secretions of salt from a nodular gland at the base of the beak near the eye. This gland pumps out the excess salt consumed with the penguin's seafood diet.

ABOVE:
Humboldt Penguin
Much of the western coastline of South America is arid. The naked skin around the beak of the Humboldt penguin is there to help the bird shed unwanted heat.

OPPOSITE:
Galápagos Penguin, Galápagos Islands
The most northerly penguin species is a relative of the Humboldt and African penguins. It is smaller and more lightweight, as it needs less fat in the warmer water. Their land habitats are mostly volcanic rock.

ABOVE:
Humboldt Penguin
Out at sea, this species relies on vast schools of little fish called anchoveta which forms the main component of its diet in the wild. The humboldt penguin also eats squid, shrimp and krill.

OPPOSITE TOP:
Humboldt Penguin
This South American species typically has a horseshoe of white that runs from the eye to the chin. Young members of this species, as seen here, have a darker head than the adults.

OPPOSITE BOTTOM:
Magellanic Penguins
The competition for breeding space is fierce. Up to 400,000 birds will crowd into traditional breeding grounds on the coasts of Chile and Argentina. The males arrive on the breeding ground first to fight each other for the best nesting spots.

Magellanic Penguins
This South American species is identified by the double stripe across the breast. It belongs to the genus *Spheniscus*, which contains three other banded penguins: African, Galápagos and Humboldt species.

Magellanic Penguin, Falkland Islands
In common with other banded penguins, the Magellanic penguin communicates with a braying call. Highly monogamous, a mate will use the calls to locate and identify his or her mate. Another kind of call is to keep non-mating penguins away. Out at sea, the penguins also call to each other to keep the hunting group together as they range far and wide.

BELOW:
Magellanic Penguins
This species was named after the Portuguese explorer Ferdinand Magellan, who encountered them during his circumnavigation of the globe in the 1520s. The birds live around the southern coastline of South America, especially Cape Horn and its many nearby islands.

RIGHT:
Magellanic Penguins, Argentina
This species of penguin builds a nest in a sheltered spot, often under a bush. However, on this arid, sandy coast, the penguins have dug burrows for their eggs.

TOP LEFT:

Northern Rockhopper Penguin
The relationship between the various rockhopper penguins, is open to debate. The northern species is the most endangered, living on and around a few islands in the South Atlantic and southern Indian Ocean.

MIDDLE LEFT:

Eastern Rockhopper Penguins
This penguin is regarded as a subspecies of the southern rockhopper, mostly centred on the sub-Antarctic island of the Indian and Pacific oceans south of Australia and New Zealand.

BOTTOM LEFT:

Southern Rockhopper Penguin
Once regarded as a single species with the northern rockhopper, these birds are now regarded as a separate species.

OPPOSITE:

Western Rockhopper Penguins, Falkland Islands
This subspecies of southern rockhopper lives on and around the Falklands. The difference between subspecies is hard to detect. There are some distinctions in the lengths and volume of the yellow crests, but the reason to treat the birds as discrete groups is that their breeding populations are so isolated they can never mix.

LEFT:
Rockhopper Penguin, Falkland Islands
As the name suggests, the rockhoppers choose rocky shorelines to build their nests. They make warm and dry nests from twigs and grasses

BELOW:
Rockhopper Penguin, Falkland Islands
While other species might waddle around or toboggan over obstacles, this species and similar crested penguins hop, leap and jump up and down the rocky shoreline..

LEFT:
Southern Rockhopper Penguin, Falkland Islands
The rockhoppers are highly social penguins. The yellow 'eyebrows' are deployed during conflicts. With a shake of the head, the crest stands up fully.

OVERLEAF:
Northern Rockhopper Penguins, Nightingale Island
In 2011, an oil spill in the South Atlantic inundated a breeding colony of northern rockhoppers on Nightingale Island. Conservationists moved the penguins en mass to Tristan da Cunha for cleaning.

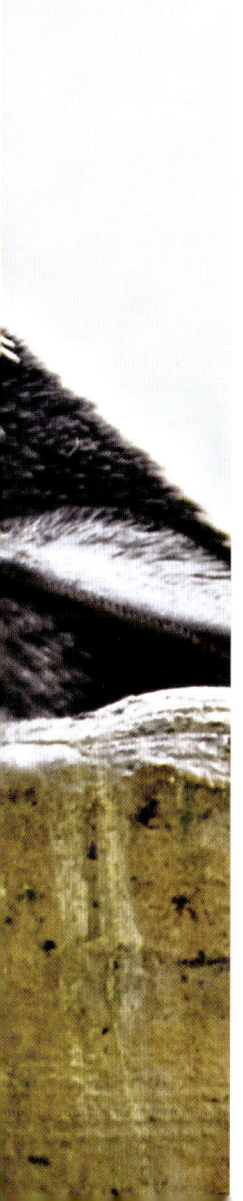

LEFT:
Rockhopper Penguin
Another of the crested penguins in the genus *Eudyptes*, rockhoppers live on the islands in every part of the Southern Ocean, preferring rocky shorelines.

ABOVE TOP AND BOTTOM:
Macaroni Penguins
This striking species was named after an early 19th-century term for a young man who was judged to be somewhat of a fashion victim for his flamboyant dress, which included a feather in his hat.

Macaroni Penguins, South Georgia

This species lives on the cliffs and rocky coastlines of islands across the South Atlantic region. During the breeding season, the birds spend the day out at sea feeding, mostly on krill, and they return at dusk to spend the night on land. Outside of the breeding season the birds spend long periods out at sea in small groups.

ABOVE:
Chinstrap Penguin, Antarctica
It is easy to understand how the chinstrap penguin got its name from the narrow black band under its head shown in this photograph.

OPPOSITE:
Chinstrap Penguin, South Georgia
The species lives on the Antarctic Peninsula, the most northerly point of the continent, and also roosts on the sub-Antarctic islands.

Chinstrap Penguins, Antarctica
As well as living on floating ice platforms outside of the breeding season, the chinstraps form large breeding colonies numbering up to 10 million birds.

Chinstrap Penguin, Antarctic Peninsula
A thick layer of blubber – a form of vasculated fat – keeps the penguin warm in the polar climate, and allows the bird to slide on their belly when crossing ice. Also known as the stonebreaker penguin, members of this species use their beaks to collect round pebbles for building a bowl-shaped nesting platform.

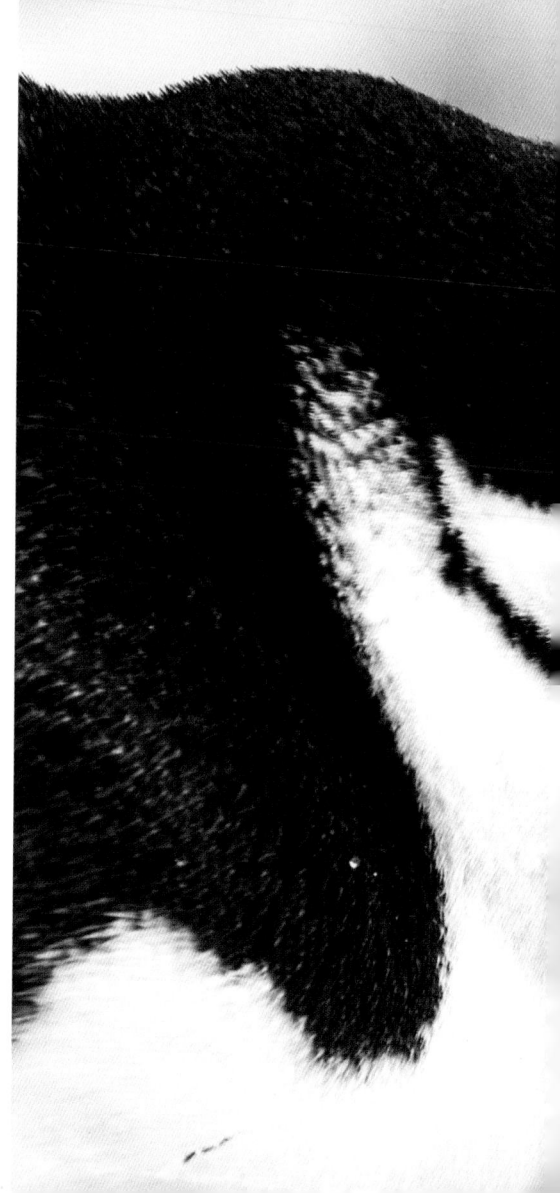

RIGHT:
Snares Penguins, Snares Islands
This species of crested penguin, a relative of the rockhoppers and royal penguin, is found only on the Snares Islands, an archipelago off the southern coast of New Zealand's South Island.

OVERLEAF:
Snares Penguins
There are an estimated 20,000 members of this species, all of whom return to the Snares Islands in summer during the breeding season – although a few stragglers have been known to end up on the New Zealand mainland.

LEFT:
Snares Penguins, Snares Islands
The Snares are the only New Zealand territories that have no introduced mammals, which allows penguins and other endemic species to thrive. The islands remain closed to the public to protect the wildlife.

OVERLEAF:
Adélie Penguins, Ross Sea, Antarctica
A species found along the entire coast of the Antarctic continent, Adélie penguins spend a large part of the year out at sea floating on ice platforms that gather around the edge of Antarctica.

PREVIOUS PAGES:

Adélie Penguins, Antarctica
Most penguin species seldom stray below the Antarctic Circle. These adélie penguins, aboard an iceberg close to the Antarctic, are exceptions.

ABOVE:

Adélie Penguins, Western Antarctica
The adélie penguin got its name from French explorer Jules Dumont d'Urville. He was the first to sight the continent in 1840, and he named the land he saw after his wife, Adéle.

OPPOSITE:

Adélie Penguin, Antarctica
This species lives all around Antarctica, breeding on the mainland and nearby islands in summer, before heading further north when winter hits. The birds will migrate about 13,000km (8,125 miles) in a year, back and forth between hunting and breeding grounds.

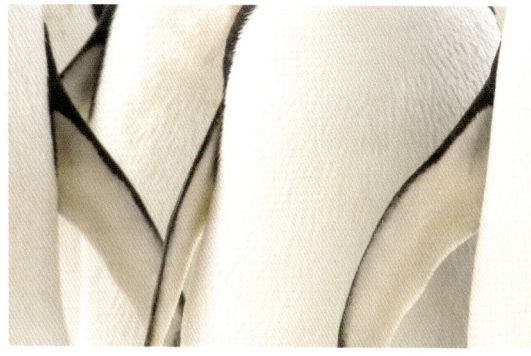

LEFT:

Adélie Penguin, Antarctica
This medium-sized species has a distinctive white ring around its brown eye, and feathers halfway up the beak. Adélie penguins rest on ice for only a short period, spending 90 percent of their time foraging for food at sea.

ABOVE TOP AND BOTTOM:

Adélie Penguin
The adélie penguin looks starkly different front and back, with white breast plumage and black backs. The birds are renowned for fighting off rivals and predators with a fierce slap with the flippers.

Adélie Penguins, Antarctica
Adélie penguins cross the ice. This species of penguin gathers in early summer to breed, forming colonies that can swell to 1.5 million individuals.

RIGHT TOP, MIDDLE AND BOTTOM:
Adélie Penguins, Antarctica
The quickest way for a penguin to cross the frozen ground of Antarctica is to slide along on the belly, pushing with the feet. This way of moving is called tobogganing.

OPPOSITE TOP:
Adélie Penguins, Antarctica
Diving in from an iceberg is a risky business. No penguin wants to be the first in case a leopard seal is lurking unseen in the water beneath. As the crowd above the water gathers, the birds at the front are at increased risk of being pushed in. The remaining penguins will look to see if this hapless bird survives.

OPPOSITE BOTTOM:
Adélie Penguins, Antarctica
This species cross ice fields on mainland Antarctica to gather at traditional breeding areas that are ice-free in summer. These penguins appear to be skating.

ABOVE TOP AND BOTTOM:
Fiordland Penguins, South Island, New Zealand
This crested penguin species lives along the southwestern shore of New Zealand's South Island, a coastline known as Fiordland due to the way it has been carved out by ancient glaciers.

RIGHT:
Fiordland Penguins
The penguins nest in burrows under rocks and tree roots in the coastal forests of the region, and follow well-trodden paths between the nests and the sea.

Fiordland Penguin
A Fiordland penguin porpoises through the water. Unlike many penguin species, the Fiordland species spends the summer at sea and returns to land to breed in winter. Little is known about what they eat, but the diet is thought to rely heavily on squid, prawns and, of course, fish.

OPPOSITE:
Royal Penguins, Macquarie Island, Australia
This species is found on and around Macquarie Island, an Australian territory located halfway between New Zealand and Antarctica.

RIGHT AND BELOW:
Royal Penguins
Despite the name, this species is not a relative of the king penguin. Instead it is a member of *Eudyptes*, the crested penguin genus, and was once seen as a subspecies of the macaroni penguin.

Royal Penguins, Macquarie Island, Australia

It is estimated that around 850,000 royal penguins gather each year on Macquarie Island to breed. They look very similar to macaroni penguins but are about 20 percent larger than their more widespread cousins and unable to breed with them.

Penguins in Water

The penguin's bird-shaped body, lightweight and built for flight, is no hindrance to the penguin. Evolution has re-purposed the sleek, aerodynamic form into a hydrodynamic one.

The wing bones have become flattened to turn the flight surfaces into more rigid flippers that cut and slice through the water. The flippers may be used as paddles or oars in different configurations, and even flap like the wings that they were as the bird darts on long dives beneath the waves. No bird is complete without a covering of feathers, which are warm and light, but also adaptable into the stiff components needed for flight. However, these features are of no use to the penguin. Instead, a thick layer of fluffy feathers keeps the body warm in air and underwater, with the longer, outer feathers forming a smooth, streamlined surface when the bird is on the move.

The water is the source of all penguins' food, which ranges from fish and squid to tiny krill and shrimps. The birds' long tongues have rear facing bristles that help to grip and cling on to slippery marine foods. Such foods are common in ocean waters that are fed by cold polar currents, which carry a high concentration of oxygen and nutrients, and it is here that penguins thrive. Many species of penguin spend their winters in the water out at sea, only resting on land or on ice for short periods. It is only the impulse to breed that forces the penguins to spend extended time on dry land.

OPPOSITE:
Adélie Penguin
Compared to the gangly lower limbs of other birds, penguins have long bodies, short legs and big, webbed feet.

Adélie Penguins
Diving is perhaps as close to flying as penguins get. The transition from land to sea is a dangerous time, when predators will be lurking above and below the water line. The penguins waste no time.

OPPOSITE TOP:
Gentoo Penguin
Freeing itself from the constraints of land, this penguin gracefully shapes its body for life in the Antarctic waters.

OPPOSITE BOTTOM:
King Penguin, East Falkland Island
Making the most of its buoyancy, this penguin is surfing up the beach, saving energy and putting off the time it will need to walk.

ABOVE:
Adélie Penguins, Antarctica
Getting into the water is a lot easier than getting out. Penguins are universally diurnal, meaning they forage during the day.

Royal Penguin, Macquarie Island, Australia
The penguin eye is tuned to see best under water. On land it is thought that penguins are somewhat near-sighted.

Adélie Penguins, Antarctica
This species spends most of the winter feeding out at sea. They rest on floating ice, especially in autumn, when they are moulting and need to stay dry and out of the water for longer periods.

Gentoo Penguin, Falkland Islands
Even the penguin, a master swimmer, is at the mercy of the power of the waves. This gentoo penguin has been caught out while trying to save some energy and ride the surf into shore.

LEFT:
African Penguins
This species tends to commute to the foraging grounds daily, returning to land each night. Most penguins are unable to hunt at night because they rely heavily on vision to locate and catch prey.

BELOW TOP:
Humboldt Penguins
This species travels further afield and spends up to 60 hours at sea.

BELOW BOTTOM:
African Penguin, South Africa
After a hard day swimming, this African penguin roosts on rocky shorelines. Members of the species identify each other by the unique pattern of dots on its breast feathers.

PREVIOUS PAGES:
African Penguins, South Africa
Penguins are seldom alone, preferring to move in groups. On land they form a waddle of penguins, for obvious reasons, and in water they are known collectively as a raft.

RIGHT:
Adélie Penguin, Antarctica
The purpose of the porpoising behaviour, where a penguin moves in short leaps over the surface, is that it likely confuses predators gathering beneath.

OVERLEAF:
African Penguins, False Bay, South Africa
While swimming submerged, penguins flap their wings, and effectively fly through the water. The birds steer with their feet, using them like rudders.

Gentoo Penguin, Falkland Islands
Porpoising behaviour allows the bird to make short leaps into the air to take deep breaths, while continuing to swim at speed in search of a food source. When food is located, the penguins will dive and feed underwater.

RIGHT:

Galápagos Penguin, Galápagos Islands
When the penguin dives, its thick layers of feathers trap bubbles of air. Not only does this help insulate the animal and stop it losing heat to the water, but it also helps boost buoyancy.

OPPOSITE RIGHT:

Jackass Penguin, Western Cape, South Africa
The penguin's feathers are coated with oils that repel water. This keeps the water away from the skin and reduces the rate at which heat is lost from the body.

BOTTOM:

Gentoo Penguin
Antarctic waters are frequently around −2°C and seldom warmer than +2°C. A thick layer of fat under the skin provides highly effective insulation – and also works like an internal floatation device or life jacket.

King Penguins
A group of king penguins waddle through the surf. Penguins are not the steadiest walkers thanks to their short legs, but once they are in the water they have arrived in their element.

Little Penguin
The smallest penguin of all is also the only one to feed in the dark, or at least during dawn and dusk. At this time the penguins assume they are safer from predators.

King Penguins, South Georgia

When bobbing at the surface of the water, penguins look much more like other water birds. They kayak along using their fins as paddles.

ABOVE:
Gentoo Penguin
While most birds have hollow bones, penguins have solid bones, to give them negative buoyancy for diving. As it descends, the air in the penguin's feathers is pushed out, helping it to sink deeper.

OPPOSITE TOP:
King Penguins
The larger penguins can dive to below 240 metres (780 feet) and hold their breath for up to 20 minutes, although dives of around five minutes are much more the norm.

OPPOSITE BOTTOM:
Crested Penguins
Here a raft of crested penguins work together to drive shoals of fish into shallower waters, where it is easier to catch them.

LEFT:

Emperor Penguin, Antarctica
With a top swim speed of 9km/h (5.6mph), this emperor penguin leaps out of the water to reach the safety of the sea ice. Emperor penguins have strong claws that help them grip the ice.

OVERLEAF:

Gentoo Penguins
Employing the technique of jumping straight from the water with a surge from their webbed feet and flippers is the only way penguins can get up steep slopes.

Penguins on Land

As flightless birds, penguins have made a life where they have no need to take to the air. Instead they are built to live above and below the waves. However, as air breathers, like all birds, penguins remain tied to the land, not for themselves but for the survival of future generations.

The bird's egg is largely a food store of fatty yolk and protein-rich white to supply energy and building materials for the developing chick. All this is kept neatly within the hard egg shell, which is waterproof – to keep the innards within as much as to stop water soaking in from without – but the shell is also permeable to the air. A small air space forms at one end (leading to those arguments about which end to crack a boiled egg) which supplies oxygen for the developing chick. As a result water birds, like penguins, cannot take their eggs out to sea and under the water. The chick would soon drown.

Penguins therefore face a choice: to breed or to eat. To incubate an egg and raise a chick means to spend considerable amounts of time away from their food sources, and for most species that also means finding ways to contend with the harsh realities of polar weather. As a result penguins have various adaptations for life on land, despite appearing to be clumsy and laboured when out of the water. For example, the feet are built for standing on snow for weeks on end, and their robust skeleton makes them sturdy enough to withstand the rough-and-tumble of life on rock and ice.

OPPOSITE:
King Penguins, South Georgia
Penguins always prefer safety in numbers on land and gather together in large rookeries to breed. King penguins will form groups of more than 30,000 birds.

Emperor Penguins, Antarctica
In the polar spring, emperor penguins divide their time between the secluded breeding grounds and frantic feeding trips at sea. Here, females freshly fattened from a winter at sea have relieved the near-starved fathers from duty.

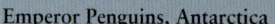

PREVIOUS PAGES:
Chinstrap Penguins
The male penguins get to the breeding ground first in spring and compete for the best places to set up a nest before the females arrive. Competition for nesting places and resources can lead to fights among males.

RIGHT:
Adelie Penguins, Antarctica
Adélie penguins gather on the ice pack in the polar autumn to shed summer feathers and grow a new plumage fit for a winter at sea.

OVERLEAF:
King Penguins, South Georgia
By the summer breeding season, the cold conditions have passed, and the penguins gather on the ice-free land. South Georgia is home to an estimated 100,000 king penguins.

Gentoo Penguins, Falkland Islands

A penguin chick chases its parent in the hope of getting more food, which the parent provides regurgitated through its beak. Soon the chick will set out into the ocean to catch its own meals.

LEFT TOP:
Emperor Penguins
Tobogganing, where the penguin slides along on its belly over the slippery ice, is a much more efficient way of moving than walking for this creature.

OPPOSITE LEFT BELOW:
Emperor Penguins, Antarctica
A group of emperor penguins slide on their bellies. These big penguins have to cover long distances to reach their breeding grounds in autumn, as the sea ice extends out into the ocean. Some emperor penguins traverse 120km (75 miles) of ice.

BELOW:
King Penguin, South Georgia
Tobogganing involves making use of a natural incline and pushing with the feet. The flippers provide balance and are used to correct the course.

Emperor Penguin, Antarctica
The sliding action is helped by the penguin's thick layer of fat that has been laid down during the autumn feeding season, and which will sustain this bird through 60 days of darkness that is the Antarctic midwinter.

Gentoo Penguins and Chinstrap Penguins, Elephant Island
It is common for penguins of two or more species to cram together on the coastlines of the sub-Antarctic islands, where space can be limited.

LEFT:

King Penguin with a Southern Elephant Seal
Penguins often share their island homes with other sea creatures, such as this southern elephant seal, who is resting before foraging.

BELOW:

King Penguins with Fur Seals
Sea lions and seals are generally good neighbours for penguins, although larger species, most notably the leopard seal, are fearsome predators of penguins.

King Penguins
Penguins frequently size each other up, quite literally in the case of king penguins. Conflicts over space and mates are often solved by the males stretching to their full height. The taller one wins.

Chinstrap Penguins, Deception Island, Antarctica

This species of penguin build nests from stones that create a frost-free platform above the ground. Members of the colony are in near constant contact using calls and visual signals, made with the head and flippers, mostly to ensure that neighbours keep their distance. If a rival gets too close, then the defender will charge them, head bowed.

Magellanic Penguin, Valdes Peninsula, Argentina

This South American penguin digs a burrow for its eggs – or more likely occupies and renovates a space vacated the previous year. In rocky areas, the breeding pairs make do with a nook on the cliffside. Once the eggs are laid – normally two – the parents will share the responsibility of incubating them for 40 days.

158

LEFT:
Emperor Penguins, Antarctica
Unlike other penguins, this species rarely stands on thawed ground. Like all penguins it has a counter-current system in the blood vessels of the feet, where warm blood from the body reheats the cold blood coming from the feet before it goes back into the body. Similarly the cold blood chills the warm blood entering the feet so it does not heat them up enough to melt the ice. In the sub-zero conditions melting ice would refreeze immediately, probably rooting the penguin to the spot.

BELOW:
King Penguins
This species breeds on barren rocky beaches with few nest-building materials. They will not scrape a nest into the ground, but brood eggs on their feet to ensure they are insulated from ground frosts.

King Penguins, South Georgia
When the weather turns bad, the penguins will huddle together. The birds around the outside will try to reach the middle and in so doing each bird takes its turn on the periphery and shields the rest from the wind.

Gentoo Penguins, Falkland Islands
When food is scarce, gentoo chicks must chase their parents to get enough to eat. It is a case of winner takes all, and this mechanism ensures that the strongest chick invariably comes out on top.

RIGHT:

Macaroni Penguins, South Georgia
This species of penguin forms enormous breeding colonies of more than a million birds. Sailors report being able to smell the rookeries from many miles out to sea.

OVERLEAF:

Snares Penguins, Snares Island, New Zealand
Uniquely isolated in a unique habitat, the Snares penguins nest among the roots of trees (which are giant relatives of daisies). The rookeries steadily erode the plant cover, so the penguins shift to new areas, allowing the trees to regrow.

Penguin Family

Family life is very much part of a penguin's life cycle, as the birds gather in vast colonies, often millions strong, that form on beaches, ice fields, forests and rocky shorelines that have been used generation after generation for many thousands of years. The breeding grounds of some South American species have been so well attended, and for so long, that the penguins digging a nest must burrow into solid guano!

On arrival at a colony, the adults seek out their partners, listening out for their unique calls among the cacophony of shrill squawks and brays. Penguins often form strong pair bonds with their mates that persist years after year, from breeding season to season. That tight-knit mother and father team will be needed.

Prior to the summer breeding season, the penguins will feed extensively, making use of the long days of the polar summer to build up reserves of fat. After the eggs have been laid – most species will lay two, but the king and emperor penguins, the biggest species, habitually lay just one – at least one parent will have to stay with their offspring and undergo a period of fasting. This can be anything from a few days to about four months.

The males generally arrive first on land. The breeding time window is short for many species, who must mate, lay eggs, incubate them and then feed the chicks until they are ready to take to the sea – and do that all before the winter arrives. If the male partner is late, his partner will not wait for him and soon move on to another mate.

OPPOSITE:
Emperor Penguins
An emperor penguin family stands together to survive a winter in Antarctica; this is the harshest habitat on Earth.

Magellanic Penguins, Punta Tombo, Chubut, Argentina

To attract a mate, the male gives a loud braying call. An interested female approaches and the male will strut around her in a circle before sealing their bond by patting her with his flippers. The new pair then preen each other, something they will do repeatedly each time they meet.

OPPOSITE:
Magellanic Penguin, Chile
Male magellanic penguins fight over the best nesting sites under bushes near the shore, and then renovate them before the female lays hers eggs.

TOP LEFT:
Galápagos Penguin, Galápagos Islands
This equatorial species sets up a nest in a rocky crevice or a burrow dug into the soft volcanic sands, choosing a location set back from the high water line. The nest contains two eggs, laid four days apart.

BOTTOM LEFT:
Gentoo Penguin, Falkland Islands
Both gentoo penguin parents work together to build a nest of stones and sticks on the ground. The female lays two eggs, and both parents share the incubation.

Rockhopper Penguins, Dunbar Island, Falkland Islands
Rockhoppers gather in vast crowds and as a result they are prone to aggressive behaviour. To avoid unnecessary conflicts the rest of the time, the birds move around the colony with their heads held downward to avoid any provocative eye contact.

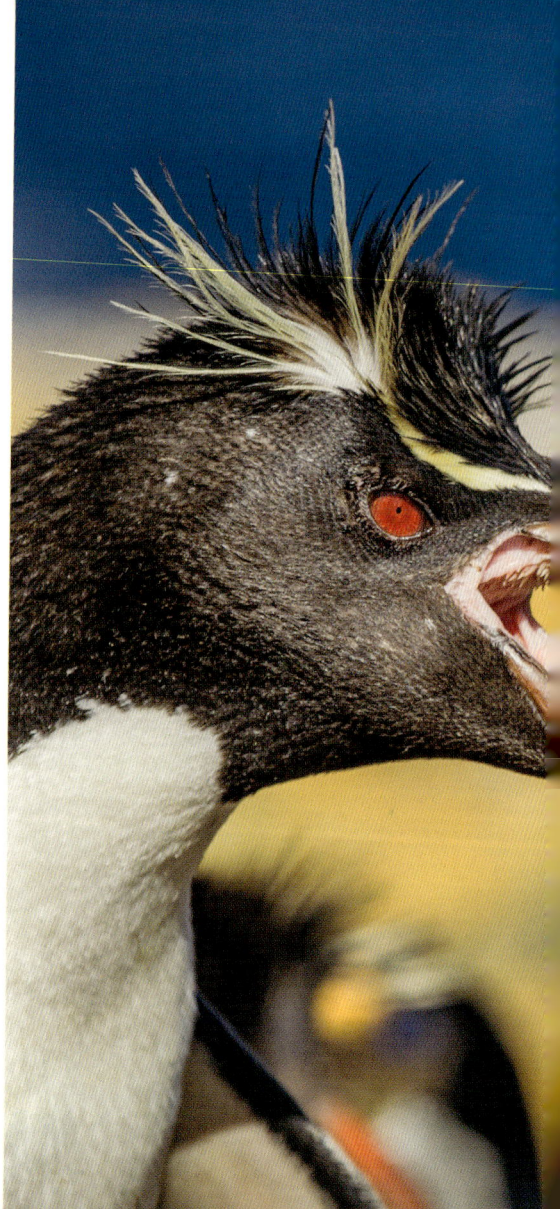

RIGHT:
Emperor Penguins, Antarctica

After laying her single egg, a female transfers it to her mate's feet. The handover must happen quickly to prevent the egg freezing in the plunging temperatures. If the egg rolls off down a slope, the waddling birds will struggle to retrieve it, and so the egg is generally abandoned.

OVERLEAF:
King Penguin, Falkland Islands

Just like its close relative, the emperor penguin, the male king penguin will keep his single egg off the ice by resting it on its feet, and incubates it under a paunch of flabby feathers that flops over the top.

ABOVE:
King Penguins, Falkland Islands
King penguins form a strong pair bond. The rigours of raising a chick mean that the pair breed twice every three years, taking a year off to recover.

OPPOSITE TOP:
Emperor Penguins, Antarctica
In spring, when the chicks are too large to sit on the feet, both parents leave their young to feed, with several chicks watched over by the remaining adults.

OPPOSITE BOTTOM:
Emperor Penguins, Antarctica
As the chicks grow larger and fatter, they are huddled together into a makeshift crèche and protected from the wind by a barrier of adults.

LEFT:

Emperor Penguins, Antarctica
The chicks could freeze to death at night if they are allowed to wander away from the colony. During the day the adults must protect them from marauding skuas and petrels, which have been known to eat up to a third of the chicks in one season.

OVERLEAF:

King Penguins, Saunders Island, Falkland Islands
A chick begs its parent for food. Reared in soil, not ice, the chicks of the king penguin grow a brown down which helps them stay hidden from predators.

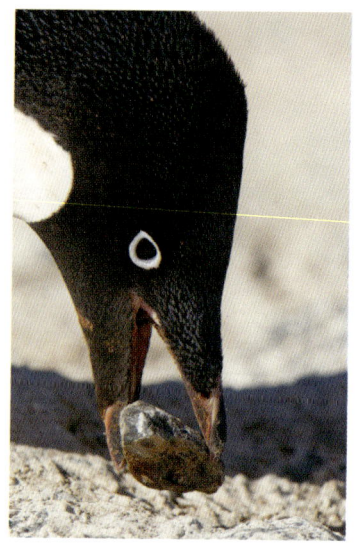

OPPOSITE, ALL IMAGES:
Adélie Penguins, Antarctica
To keep eggs off the frozen ground, adélie penguins gather stones to make a nesting platform.

LEFT:
Gentoo Penguin, French Southern and Antarctic Lands
Ground-nesting gentoos lay eggs on platforms of twigs and pebbles.

BELOW:
Magellanic Penguins, Bleaker Island, Falkland Islands
By the age of two-and-half months the chick is nearly fully grown. It sheds the babyish down feathers to reveal the adult plumage.

Gentoo Penguin, Saunders Island, Falkland Islands

The crowded gentoo colony rapidly fills up with orange droppings, and breeding pairs move their eggs to a new nest in a cleaner spot every three days or so.

ALL PHOTOGRAPHS:
Gentoo Penguins, Falkland Islands
Both parents care for the young equally. It takes a little over a month for the eggs to hatch and the chicks will take nearly five months to fledge. The parents regurgitate a portion of their own meals for the chicks to eat. When food from the sea is scarce, the chicks will fight for each meal; generally only one survives.

Gentoo Penguins, Falkland Islands

This species of penguin is among the most faithful, and will maintain pair bonds over many years, even as breeding grounds and nesting sites are frequently shifting locations. Most of the penguins stay close to their familial breeding grounds all year, only making short foraging trips further afield.

Penguin Chicks

It often requires a double-take to identify a penguin chick. Spending their formative days on land, as they do, the chick's plumage is devoted to solving a different set of challenges than that of its ocean-going parents. The most obvious difference is that the chick lacks the black-and-white counter-shading of the adult. When newly hatched, the chick is more or less naked, but after growing rapidly and growing large enough to demand solid foods, the chick must prepare to stay ashore by itself while both its parents head out to sea to find fish and other tasty titbits. In response, the chick grows a thick coat of fluffy down feathers. This swaddles the bird by trapping air against its skin, helping it keep out the cold wind and maintain its body temperature.

This plumage stands the chick in good stead as it scoffs meals brought by its parents and builds up a thick layer of subcutaneous fats. When the downy feathers get wet, they will lose their powers of insulation, so a coat of blubber is needed before the young bird can transition to life at sea. Indeed, unseasonal rainstorms hitting the breeding colony can be deadly as bedraggled chicks die of the cold.

As the day approaches when the chick will head out to join its parents at sea, the fluffy down is moulted away, revealing the sleek penguin suit of well-oiled feathers beneath. The chick will not return to its breeding grounds for a couple of years at least. When it does, it will be to care for its own chick.

LEFT:
King Penguin Chick, Falkland Islands
Watching and waiting, the king penguin chick takes more than a year to fledge. It has to wait for a month or longer between feeds from a parent and as it waits it can lose up to half its body weight under its plump fluffy exterior.

Emperor Penguin Chicks, Antarctica
Despite being bigger than its cousin the king penguin, the chicks of the emperor penguin cannot wait so long before taking to the sea. Five months after hatching they will head out to sea with their parents, albeit still to develop their adult plumage.

RIGHT:
Emperor Penguins, Antarctica
An emperor penguin chick's first meal is a milk-like secretion from its father's oesophagus. The father has 10 days' supply of this goop before he is too emaciated to continue. Then the chick is fed with regurgitated food from the mother, who has returned from her winter foraging.

OPPOSITE:
Gentoo Penguins, Falkland Islands
After hatching, the gentoo chicks are almost helpless and have only a scant covering of feathers. They stay at the nest for nearly three months before growing strong enough to make short forays to explore their surroundings.

King Penguin Chick, Macquarie Island, Australia
At about the age of 14 months, the king penguin chick begins to shed its baby feathers. Every baby feather is replaced with a new adult one.

FAR LEFT:

Emperor Penguins
During the spring, the emperor penguin parents are constantly taking it in turns to travel to the sea, feed as quickly as possible and return to the breeding ground to alleviate their partner and feed the hungry chick.

TOP LEFT:

Emperor Penguin Chick, Antarctica
A chick calls to its parent for food. After hatching, the chick weighs around 400 grams (14oz). It will have to grow quickly to around 12kg (26.5lbs) before it is ready to leave the land and make a life at sea.

BOTTOM LEFT:

Emperor Penguins, Antarctica
The little chick cannot stay out on the ice for long without getting dangerously cold. It clambers back onto Mum or Dad's feet to warm up again.

King Penguins, Falkland Islands

An adult stands guard among a crèche of king penguin chicks, which have spent their first year on land, mostly waiting for their parents to return with the next meal. If the food does not arrive soon enough then the chicks, unable to enter the water to feed themselves, will starve. The few adults that are on land resting before the next foraging trip also work hard to protect the chicks from the other great danger: being killed and eaten by a skua or other tough seabird.

King Penguin Chick, Volunteer Beach, East Falkland Island
A king penguin chick rests on its parent's feet. Just after hatching the king penguin chick is almost naked, featherless and entirely helpless.

OPPOSITE:

Gentoo Penguin Chick, Falkland Islands
After making early swims in the ocean, and while still learning to hunt for themselves, juvenile gentoos will return to the breeding ground and take extra meals from their parents.

ABOVE:

King Penguin Chicks
The straggly down feathers of the king penguin chicks keep them warm on the windswept beach as they wait for the next supply run from their parents.

OVERLEAF:

Gentoo Penguin Chicks, Falkland Islands
The chicks of gentoo penguins fledge at around the age of 70 days and start to make short swimming trips. They will reach full independence after about a hundred days.

OPPOSITE:

King Penguin Chick, Falkland Islands
The fluffy feathers of a chick would be a severe hindrance in the water, so they must be moulted. When wet they would hamper the hydrodynamics, creating drag as the bird moved through the water.

LEFT:

Adélie Penguin Chick, South Orkney Islands
Water takes away body heat 40 times faster than air, so to prepare for life as a water bird, the chick renews its plumage with well-oiled, slick feathers that traps air against the skin better.

OVERLEAF:

Gentoo Penguin Chicks, Antarctica
These chicks appear to be calling to their parents for food. Most penguin chicks have the summer to grow. Come the autumn, when the weather turns colder, it is time for them to head out to the relative warmth and plenty of the open sea.

OPPOSITE:
Gentoo Penguins, Falkland Islands
The first few weeks of life will be tough for this pair. If they are lucky enough to be born in a time of plenty then they will both make it to adulthood. But, it is likely that their parents will not be able find enough food for both of them, and they will become locked in a competition for food that only one will win.

BELOW:
Gentoo Penguin Chick, Falkland Islands
Nearly two-and-half months after hatching, a gentoo chick sheds its down and becomes transformed into an adult. The new feathers will be waterproof and allow the penguin to survive in the sea.

Emperor Penguin Chicks, Queen Maud Land, Antarctica

This gang of chicks is bunching together to fend off the Antarctic chill. After receiving one-on-one care from either parent for most of two months after hatching, the chicks spend the next hundred days with both parents away for long periods.

Emperor Penguins, Adélie Land, Antarctica
The adult male emperor penguin performs one of the great feats of parenthood. After walking and tobogganing for days to reach the breeding site, the bird sits out the −70°C winter, keeping the egg warm all the while, and feeds the chick its first meals. And it does this without having a single meal itself. Having lost half its body weight, it will eventually reach the sea to feed again 115 days after departing the autumn before.

Emperor Penguin Chick, Antarctica
An emperor penguin chick sits on its parent's feet. The emperor penguin is miraculous in many ways. Its greatest feat is that of survival. Thanks to the dedication and teamwork of its parents, this is the only land animal to be born in the midst of the Antarctic winter.

Picture Credits

Alamy: 24/25 bottom (Image Broker), 26 (Cavan), 28/29 (Steve Sadler Images), 30/31 (Olga Khoroshunova), 32 bottom (Horizons WWP), 34/35 (Fritz Polking), 47 bottom (blickwinkel), 60/61 (Agami Photo Agency), 68/69 (Steve Bloom Images), 72/73 (Nature Picture Library), 90 top (Image Broker), 91 (National Geographic Image Collection), 102 top (David Graham), 116/117 (Nature Picture Library), 129 bottom (Richard Robinson), 130/131 (Steve Bloom Images), 150 (robertharding), 152/153 (David Osborn), 158/159 top (Nature Picture Library), 166/167 (Frans Lanting Studio), 181 bottom (Nature Picture Library), 203 bottom (Steve Bloom Images), 214/215 (robertharding), 216 (National Geographic Image Collection)

Dreamstime: 7 (Ondrej Prosicky), 17 (Sergey Uryadnikov), 32 top (Filip Fuxa), 38/39 (Vladimir Seliverstov), 42/43 (Sylvia Adams), 44 (Phichak Limprasutr), 45 (Pablo Hidalgo), 46 (Gueret Pascale), 47 top (Irakite), 48/49 & 52 (Ondrej Prosicky), 54 top (Wrangel), 54 middle (Chuckaitch), 54 bottom & 55 (Dalia Kvedaraite), 62 (Leopold Brix), 63 top (Legacy Images), 74/75 (Agami Photo Agency), 80/81 (Andrea Basile), 86/87 (Staphy), 94 & 95 bottom (Agami Photo Agency), 98 (Andreanita), 110 (Cathy Withers), 111 top (Theodor Bunica), 111 bottom (Anyaberkut), 118/119 (Dalia Kvedaraite), 120/121 (Waravit Vijitpanya), 122/123 (Andrea Basile), 124/125 (Martingraf), 128 (Rafael Ben Ari), 138/139 (Igor Gushchin), 140/141 (Jonathan R. Green), 146/147 top (Tenedos), 147 bottom & 160/161 (Willtu), 162/163 (Fieldwork), 173 (Ondrej Prosicky), 181 top (Vladimir Seliverstov), 192/193 (Theroff 97), 196/197 (Vladimir Seliverstov), 204/205 (Jeremy Richards), 209 (Jocrebbin)

FLPA: 13 bottom (James Stone), 16 (Juergen & Christine Sohns), 27 bottom, 70/71 & 102 bottom (Kevin Elsby), 108/109 (Pierre Lobel/Biosphoto), 114/115 (Antoine Dervaux/Biosphoto), 120 top (Franco Banfi/Biosphoto), 121 top (A@Biosphoto), 126/127 (Otto Plantema), 134 & 142/143 (Kevin Elsby), 144/145 (Frans Lanting), 186 top right (Jue rgen & Christine Sohns), 186 bottom (Roger Tidman), 188/189 & 200 (A@Biosphoto)

FLPA/Biosphoto/Samuel Blanc: 6, 37 both, 176/177, 220/221

FLPA/Minden Pictures: 12 (Suzi Eszterhas), 13 top (Andrew Peacock), 36 (Stefan Christmann), 57 (Michael Milicia, BIA), 67 (Otto Plantema), 82 (Yva Momatiuk & John Eastcott), 88 top & bottom (Jan Vermeer), 88 middle (Kevin Schafer), 89 top (Otto Plantema), 89 bottom (Jan Vermeer), 103 (Suzi Eszterhas), 106/107 (Colin Monteath), 184/185 (Heike Odermatt), 190 bottom left (Franka Slothouber, NIS), 190 bottom right (Otto Plantema, Buiten-Beeld), 213 (Suzi Eszterhas), 218/219 (Stefan Christmann)

FLPA/Minden Pictures/Augustin Esmoris: 10/11, 27 top, 180, 190 top, 194

FLPA/ Minden Pictures/J-L Klein & M-L Hubert: 8, 50/51, 56, 66, 83, 148/149, 182/183, 198, 199, 203 top, 212, 222/223

FLPA/Minden Pictures/Tui De Roy: 13 middle, 14/15, 58/59, 78/79, 84, 85 bottom, 104/105, 173 top, 174/175, 178/179, 186 top left, 191 all, 206/207, 208

Getty Images: 23 (Galaxiid), 53 (John Brown), 90 bottom (Doug Gimesy), 92/93 (Ian Waldie), 132/133 (David Merron), 164/165 (David Tipling), 168 (David Merron Photography), 170/171 (Peter Giovannini), 172 (Nora De Angelli), 187 top (De Agostini Picture Library)

Shutterstock: 18/19 (Cameris), 20/21 (Coulanges), 22 (Agami Photo Agency), 24/25 top (Foto Mous), 33 (Michael Smith ITWP), 40/41 (Tria Adha), 63 bottom (Anton Rodionov), 64/65 (Knelson20), 76/77 (Janelle Lugge), 85 top (NaturesMomentsuk), 95 top (Keith Michael Taylor), 96/97 (BMJ), 100/101 (NaturesMomentsuk), 112/113 (Sergey Uryadnikov), 129 top (Tack-Ma), 136/137 (Edward Gault), 146 bottom (BMJ), 151 top (Jared Cohn), 151 bottom (Bruce Wilson Photography), 154/155 (Earth Trotter Photos), 156/157 (Joost Van Uffelen), 158/159 bottom (online express), 187 bottom (Jeremy Richards), 202 (Mario_Hoppmann), 210/211 (Johnny Giese), 217 (Andreea Dragomir)